KILL THE HEAT FIRST BEFORE IT KILLS YOU

SURVIVING THE 2023 HEAT WAVE IN AMERICA AND EUROPE

PROF. EBONY M. BINGHAM

2

TABLE OF CONTENTS

5

1

UNDERSTANDING HEAT WAVES

WHAT IS A HEAT WAVE?

A heat wave is a prolonged period of excessively hot weather that occurs within a specific region. It is characterized by significantly higher temperatures than the average climate conditions for that area during a particular season. Heat waves are often accompanied by high humidity levels, which can exacerbate the discomfort and health risks associated with the extreme heat.

Heat waves typically last for several days or even weeks, and their intensity can vary depending on the geographical location and the time of year. While heat waves are commonly associated with the summer months, they can occur at any time, even in spring or fall.

During a heat wave, the temperature rises to abnormally high levels, making it challenging for the body to cool down through its usual mechanisms, such as sweating. This can lead to a range of health problems, particularly for vulnerable populations such as the elderly, young children, pregnant women, and individuals with pre-existing medical conditions.

THE IMPACT OF HEAT WAVES ON HEALTH

Heat waves have a significant impact on human health, with potentially severe consequences. The excessive heat and prolonged exposure can strain the body's ability to regulate its internal temperature, leading to a variety of heat-related illnesses. Some of the health risks associated with heat waves include:

Heat Exhaustion

Heat exhaustion is a condition that occurs when the body becomes dehydrated and is unable to cool down efficiently. Symptoms may include heavy sweating, weakness, dizziness, nausea, headache, and muscle cramps. If not addressed early, heat exhaustion can lead to heat stroke.

Heat Stroke

Heat stroke is a life-threatening condition that requires urgent medical attention. It occurs when the body's internal temperature reaches a dangerous level, typically above 104°F (40°C). Symptoms include confusion, disorientation, rapid heartbeat, throbbing headache, dry and hot skin, and loss of consciousness. If heat stroke is not treated right away, it might cause organ failure and death.

Dehydration

Prolonged exposure to high temperatures without adequate fluid intake can lead to dehydration. Dehydration can cause fatigue, dizziness, dry mouth, decreased urine output, and in severe cases, confusion and unconsciousness.

Respiratory and Cardiovascular Issues

Heat waves can worsen existing respiratory and cardiovascular conditions, such as asthma, chronic obstructive pulmonary

11

disease (COPD), and heart diseases. The heat and poor air quality during heat waves can trigger respiratory distress and cardiovascular complications.

CLIMATE CHANGE AND HEAT WAVES

Climate change plays a significant role in the increasing frequency and severity of heat waves. The accumulation of greenhouse gases in the Earth's atmosphere, primarily from human activities such as burning fossil fuels, leads to a warming effect known as global warming. This rise in global temperatures contributes to the occurrence and intensification of heat waves.

Scientific studies and climate models project that heat waves will become more frequent, longer-lasting, and more intense as the Earth's climate continues to warm. This underscores the importance of understanding the connection between climate change and heat waves to effectively address their impacts.

By recognizing the characteristics and risks associated with heat waves, as well as the influence of climate change on their occurrence, we can take proactive measures to protect ourselves and our communities. The subsequent chapters of this book will provide practical strategies and guidance for surviving and mitigating the effects of heat waves, ensuring our well-being during these extreme weather events.

2

PREPARING FOR THE HEAT WAVE

Preparing for a heat wave is crucial to ensure your safety and well-being during extreme hot weather. By taking proactive measures, you can minimize the risks and maximize your ability to cope with the high temperatures. The following sections provide valuable information on how to prepare effectively:

Monitoring Weather Forecasts and Heat Advisories

By often checking weather forecasts, you may stay informed about the local weather conditions. Pay attention to heat advisories and warnings issued by local authorities and meteorological agencies. These alerts provide valuable information about the expected heat index, heat wave duration, and necessary precautions. By staying updated, you can plan ahead and make informed decisions regarding outdoor activities, hydration, and cooling strategies.

Creating an Emergency Heat Wave Plan

Developing an emergency heat wave plan is essential to ensure your safety and that of your loved ones. When building your plan, take into account the following factors:

Establishing Communication

Designate a communication plan with your family and friends to stay connected during a heat wave. Share emergency contact numbers and establish a meeting point in case of evacuation.

Identifying Cooling Centers and Shelters

Locate nearby cooling centers, community centers, or public facilities that offer air-conditioned spaces during heat waves.

Identify emergency shelters in case you need to evacuate your home due to extreme heat or power outages.

Stocking Up on Essential Supplies

Prepare by stocking up on essential supplies before the heat wave hits. Consider the following:

Water: Ensure an ample supply of drinking water for both you and your pets.

Store water in clean, sealed containers and consider having a water filtration system or water purification tablets as a backup.

Food: Have non-perishable food items available that require little to no cooking.

Opt for foods with high water content, such as fruits and vegetables, to help maintain hydration.

Medications and First Aid: Ensure an adequate supply of prescription medications.

Include a first aid kit with supplies for treating heat-related illnesses.

Emergency Equipment: Have battery-powered fans, portable air conditioners, or cooling towels available as alternatives to traditional air conditioning. Keep a battery-operated or hand-cranked radio for accessing emergency broadcasts.

3

PREPARING YOUR HOME FOR EXTREME HEAT

Take steps to prepare your home for the heat wave to create a cooler and more comfortable environment:

Insulation and Shade

Insulate your home to prevent heat from entering and cool air from escaping.

Install blinds, curtains, or shades to block direct sunlight and reduce heat gain.

Ventilation

Ensure proper ventilation by using fans, opening windows during cooler periods, and utilizing cross-ventilation techniques.

Consider installing attic and whole-house fans to enhance airflow.

Air Conditioning

Maintain and service your air conditioning system before the heat wave begins. Set your thermostat at a comfortable temperature and use energy-efficient cooling practices.

Cool Room Preparation

Designate a cool room in your home where you can seek refuge during the hottest times of the day. Keep this room well-ventilated and equipped with a portable air conditioner or fans.

By preparing in advance and taking these necessary precautions, you can minimize the impact of a heat wave on your daily life. Being proactive allows you to stay cool, hydrated, and safe during extreme heat conditions.

STAYING COOL INDOORS

During a heat wave, staying cool indoors is crucial to protect yourself from the excessive heat and reduce the risk of heat-related illnesses. Implementing effective cooling strategies will help maintain a comfortable and safe environment. Consider the following tips for staying cool indoors:

Using Air Conditioning Effectively

Air conditioning is one of the most effective ways to cool down indoor spaces during a heat wave. Maximize its effectiveness with these tips:

Temperature Setting: Set your thermostat to a comfortable temperature, typically between 72°F (22°C) and 78°F (26°C). Avoid lowering the temperature too much because doing so will tax the system and use more energy.

Airflow and Vents: Ensure proper airflow by keeping doors and windows closed while the air conditioner is running. Clean or replace air filters regularly to maintain optimal air quality and system efficiency. Ensure that vents and registers are unobstructed to allow efficient airflow.

Programmable Thermostats: Utilize programmable thermostats to automatically adjust the temperature settings based on your schedule. Set higher temperatures when you're away to conserve energy and lower utility bills.

Alternative Cooling Methods

If you don't have access to air conditioning or want to supplement its effectiveness, consider alternative cooling methods:

Fans: Use ceiling fans, floor fans, or portable fans to circulate air and create a cooling breeze. Place fans strategically to direct airflow toward you or across the room for better circulation.

Natural Ventilation: Open windows and doors during cooler periods, such as early mornings or late evenings, to allow fresh air into your home. Utilize cross-ventilation by opening windows on opposite sides of your home to create a cooling breeze.

Window Coverings: Install light-colored blinds, curtains, or shades to block direct sunlight and reduce heat gain.

Consider reflective window films or solar shades to further minimize heat transfer.

Cooling Towels and Misting: Use cooling towels or misting devices to lower your body temperature and provide temporary relief from the heat. Wet a towel with cool water and place it on your neck, wrists, or forehead for a refreshing sensation.

Creating a Cool Room or Shelter

Designating a cool room or shelter within your home can provide respite from the heat. Follow these tips:

Room Selection: Choose a room on the lower level of your home, as heat rises, and lower levels tend to be cooler.

Preferably select a room with windows for natural ventilation or easy installation of portable air conditioners.

Window Treatments: Use blackout curtains or thermal shades in the cool room to minimize heat penetration. Apply reflective window films to reduce solar heat gain while maintaining visibility.

Cooling Equipment: Install a portable air conditioner or evaporative cooler in the cool room for additional cooling. Place fans strategically to improve air circulation within the room.

Hydration and Comfort: Keep cool drinks, water, and snacks readily available in the cool room. Create a comfortable seating area with lightweight, breathable furniture and cushions.

Managing Humidity and Ventilation

High humidity levels can intensify the feeling of heat. Follow these tips to manage humidity and improve ventilation

Dehumidifiers: Use dehumidifiers to remove excess moisture from the air, making it feel more comfortable. Empty and clean the dehumidifier regularly to maintain its efficiency.

Bathroom and Kitchen Ventilation: Run exhaust fans or open windows in bathrooms and kitchens to expel hot air and moisture.

Use range hoods when cooking to divert heat and humidity outside.

Avoid Heat-Generating Activities: Minimize heat-generating activities, such as using the oven or stove, during the hottest times of the day.

Opt for lighter meals that require less cooking or consider outdoor grilling instead.

By implementing these strategies, you can create a cool and comfortable indoor environment, reducing the risk of heat-related illnesses and ensuring your well-being during a heat wave. Remember to stay hydrated and monitor your body for any signs of overheating or dehydration.

4

STAYING HYDRATED

Staying properly hydrated is crucial during a heat wave to prevent dehydration and maintain your body's ability to regulate temperature. The following guidelines will help you stay hydrated and minimize the risk of heat-related illnesses:

IMPORTANCE OF HYDRATION IN HOT WEATHER

In hot weather, your body loses water more rapidly through sweating to cool down. Adequate hydration is essential to replace the lost fluids and maintain optimal body function. Benefits of staying hydrated during a heat wave include:

Temperature Regulation: Proper hydration helps regulate your body temperature and prevents overheating.

Enhanced Physical Performance: Well-hydrated muscles and cells improve physical performance and reduce the risk of muscle cramps or fatigue.

Cognitive Function: Hydration supports cognitive function, concentration, and mental clarity, helping you stay focused and alert.

Electrolyte Balance: Maintaining electrolyte balance is crucial for optimal cellular function and preventing imbalances that can lead to muscle cramps and weakness.

CHOOSING THE RIGHT FLUIDS

Not all fluids are equally effective for hydration. Opt for the following hydrating fluids during a heat wave:

29

Water: Water is the best choice for hydration, as it is readily available and helps replenish lost fluids. Drink water before, during, and after physical activity, even if you do not feel thirsty.

Electrolyte-rich Beverages: Sports drinks or electrolyte-enhanced beverages can be beneficial during prolonged physical exertion or in cases of excessive sweating. These drinks help replenish electrolytes lost through sweat, especially if you engage in intense exercise.

Fruit and Vegetable Juices: Natural fruit and vegetable juices, such as watermelon or cucumber juice, are not only hydrating but also provide additional nutrients and antioxidants.

Avoid juices high in added sugars, as they can contribute to dehydration.

Coconut Water: Coconut water is a natural, electrolyte-rich beverage that can help replenish fluids and electrolytes lost through sweating. It is a refreshing alternative to sports drinks, providing hydration with a mild, natural flavor.

WATER CONSERVATION TIPS

During a heat wave, water conservation is essential to ensure an adequate supply for hydration and other essential needs. Consider the following water conservation tips:

Hydrate Before Going Out: Drink plenty of water before leaving home to start the day well-hydrated.

Use Reusable Water Bottles: Carry a reusable water bottle with you to stay hydrated on the go, reducing the need for single-use plastic bottles.

Limit Water Usage: Take shorter showers and reduce water flow while bathing to conserve water. Use a bucket to collect shower or bath water for other purposes, such as flushing toilets or watering plants.

Water-Efficient Appliances: Opt for water-efficient appliances, such as low-flow showerheads and faucets, to minimize water usage.

SIGNS OF DEHYDRATION AND HEAT-RELATED ILLNESSES

It's crucial to recognize the signs of dehydration and heat-related illnesses. Be aware of the following symptoms:

Mild Dehydration: Dry mouth and throat, dark urine, fatigue, thirst, and reduced urine output.

Moderate to Severe Dehydration: Dizziness, lightheadedness, confusion, rapid heartbeat, dry and cool skin, irritability, and sunken eyes.

Heat Cramps: Muscle cramps or spasms, particularly in the legs or abdomen, often accompanied by excessive sweating.

Heat Exhaustion: Profuse sweating, weakness, fatigue, nausea, headache, dizziness, clammy skin, and rapid pulse.

Heat Stroke (Medical Emergency): High body temperature (above 104°F or 40°C), confusion, altered mental state, seizures, hot and dry skin, rapid breathing, and loss of consciousness. Call emergency services immediately. If you experience severe dehydration or symptoms of heat-related illnesses, seek medical attention promptly.

By prioritizing hydration, choosing the right fluids, conserving water, and recognizing signs of dehydration and heat-related illnesses, you can stay properly hydrated and minimize the risks associated with heat waves. Remember to drink water frequently, even if you don't feel thirsty, and monitor your body's hydration levels during periods of extreme heat.

5

DRESSING FOR THE HEAT

Choosing appropriate clothing during a heat wave is essential to stay cool, comfortable, and protected from the sun's rays. The right clothing can help regulate body temperature, promote airflow, and minimize heat absorption. Consider the following tips when dressing for the heat:

CHOOSING BREATHABLE AND LIGHTWEIGHT FABRICS

Opt for fabrics that are breathable and lightweight to facilitate airflow and allow sweat to evaporate. Here are some ideal fabric choices for hot weather:

Cotton: Cotton is a natural fiber that is breathable and allows air circulation, keeping you cool and comfortable. Choose loose-fitting cotton clothing to maximize airflow and promote sweat evaporation.

Linen

Linen is a lightweight and highly breathable fabric that absorbs moisture and dries quickly. Wearing linen clothing helps promote ventilation and allows heat to escape from your body.

Moisture-Wicking Fabrics

Look for synthetic fabrics specifically designed for moisture-wicking properties, such as polyester or nylon blends. These fabrics pull moisture away from your skin, allowing it to evaporate more quickly.

Lightweight Knits

Opt for lightweight knitted fabrics that have an open weave, such as jersey or mesh. These fabrics promote air circulation and enhance breathability.

PROTECTIVE CLOTHING AND ACCESSORIES

While it's important to wear lightweight and breathable clothing, it's equally crucial to protect yourself from the sun's harmful rays. Consider the following protective clothing and accessories:

Wide-Brimmed Hats

To protect your face, neck, and ears from harsh sunlight, put on a hat with a wide brim.

Choose hats made of breathable materials, such as straw or cotton, for better ventilation.

Sunglasses

Put on UV-protective sunglasses to shield your eyes from the sun's rays. For the best eye protection, look for sunglasses that filter both UVA and UVB radiation. Look for sunglasses that block both UVA and UVB rays for optimal eye protection.

Sun-Protective Clothing

Consider wearing clothing with built-in sun protection, such as UPF (Ultraviolet Protection Factor) garments. These clothing items are specifically designed to block harmful UV rays and offer an extra layer of defense against the sun.

Lightweight, Long-Sleeved Shirts and Pants

While it may seem counterintuitive, wearing lightweight, long-sleeved shirts and pants can offer protection from the sun while allowing airflow and preventing direct sunlight from reaching your skin. Choose loose-fitting garments to enhance ventilation and promote sweat evaporation.

FASHIONABLE AND FUNCTIONAL HEAT WAVE OUTFITS

You can stay stylish and comfortable during a heat wave with the following outfit ideas:

Sundresses and Maxi Dresses

Opt for loose-fitting, lightweight sundresses or maxi dresses made of breathable fabrics like cotton or linen. These flowy garments allow air circulation and provide a fashionable choice for hot weather.

Shorts and Tank Tops

Choose lightweight shorts and tank tops made of breathable materials for a casual and comfortable outfit. Look for moisture-wicking fabrics to keep you cool and dry.

Breathable Footwear

Wear open-toed sandals or shoes made of breathable materials like canvas or mesh. Avoid footwear that retains heat, such as heavy boots or shoes with thick soles.

Layering for Versatility

Opt for light, loose layers that can be easily removed or adjusted based on the changing temperature throughout the day. Layering allows you to adapt to varying levels of heat and maintain comfort.

Remember to apply sunscreen to exposed skin, regardless of the clothing you choose. Sunscreen with a high SPF (Sun Protection Factor) helps protect your skin from harmful UV rays.

By selecting breathable fabrics, incorporating protective clothing and accessories, and considering fashionable and functional outfit choices, you can dress appropriately for the heat and ensure both comfort and sun protection during a heat wave.

6

OUTDOOR ACTIVITIES AND EXERCISE

Engaging in outdoor activities and exercise during a heat wave requires special considerations to ensure your safety and well-being. While it's essential to stay active, it's equally important to take precautions and adjust your routine to minimize the risk of heat-related illnesses. Follow these guidelines for outdoor activities and exercise during hot weather:

EXERCISING SAFELY IN HOT WEATHER

Time of Day

Plan your outdoor activities and exercise during the cooler parts of the day, such as early mornings or evenings. Avoid exercising during the hottest times, typically between 10 a.m. and 4 p.m.

Hydration

Drink plenty of water before, during, and after your outdoor activities to maintain hydration. Carry a water bottle and observe regular hydration breaks.

Dress Appropriately

Wear lightweight, breathable, and moisture-wicking clothing.

To stay cool and reflect sunshine, choose light-colored clothing.

Sun Protection

Even on cloudy days, cover any exposed skin with sunscreen with a high SPF. Wear a wide-brimmed hat and sunglasses for added sun protection.

Listen to Your Body

Pay attention to your body's signals. If you start feeling dizzy, lightheaded, or excessively fatigued, take a break and find shade.

HEAT WAVE SAFETY TIPS FOR HIKING AND CAMPING

If you plan to hike or camp during a heat wave, take these additional precautions:

Check Trail Conditions

Research trail conditions and choose shaded or cooler trails whenever possible. Be aware of any trail closures or restrictions due to fire risks or extreme heat.

Start Early and Finish Early

Begin your hike early in the morning to avoid the peak heat hours. Wrap up your hike or outdoor activities before the hottest part of the day.

Plan for Shade

Plan your hike or camping trip in areas with ample shade, such as forests or areas with natural canopies. Take breaks in shaded areas to cool down and rest.

Carry Sufficient Water

Ensure you have enough water for the duration of your hike or camping trip, considering the increased fluid needs in hot weather. If water sources along the trail are available, bring a water filtration system or water purification tablets as a backup.

POOL SAFETY AND WATER RECREATION

Swimming and water recreation can provide relief during a heat wave, but safety should remain a top priority:

Supervision

Never swim alone, and ensure there is appropriate supervision for children and inexperienced swimmers.

Maintain constant vigilance when near water, even in shallow areas.

Hydration

Stay hydrated by drinking water before, during, and after swimming or engaging in water activities. Avoid consuming alcohol or caffeine, as they can contribute to dehydration.

Sun Protection

Apply waterproof sunscreen to protect your skin from the sun's harmful rays while swimming or participating in water activities. Reapply sunscreen after swimming or excessive sweating.

Water Safety Skills

Learn basic water safety skills, including swimming techniques and rescue techniques, to ensure your safety and the safety of others.

SAFETY MEASURES FOR PETS AND CHILDREN

Ensure the safety of your pets and children during outdoor activities in hot weather:

Pet Safety

Limit outdoor activities for pets during the hottest times of the day. Provide shade and fresh water for your pets at all times. Avoid walking dogs on hot pavement, as it can burn their paw pads.

Child Safety

Never leave children unattended in a vehicle, even for a short period, as cars can quickly become dangerously hot. Keep children well-hydrated and dress them in lightweight, breathable clothing.

Schedule outdoor playtime during cooler hours of the day. By following these safety guidelines and adjusting your outdoor activities and exercise routines accordingly, you can enjoy the benefits of being active while minimizing the risks associated with heat waves. Prioritize safety, stay hydrated, and listen to your body's cues to ensure a safe and enjoyable experience outdoors during hot weather.

7

COPING WITH POWER OUTAGES

During a heat wave, power outages can occur due to increased energy demand and strain on the electrical grid. Coping with power outages requires preparation and knowing how to stay safe and comfortable during extended periods without electricity. Consider the following tips for coping with power outages during a heat wave:

PREPARING FOR POWER OUTAGES

Emergency Supplies

Create an emergency kit that includes essential supplies such as flashlights, batteries, portable fans, and a battery-operated or hand-cranked radio.

Keep a stock of non-perishable food, bottled water, and necessary medications.

Battery Backup and Power Banks

Invest in battery backup systems or power banks to charge essential devices like cell phones or medical equipment. Ensure they are fully charged before the power outage occurs.

Coolers and Ice

Have coolers and ice packs on hand to keep perishable foods from spoiling during a power outage. Limit opening the refrigerator and freezer to preserve the cool temperature as much as possible.

Generator Safety

If using a generator, follow proper safety guidelines and ensure it is operated in a well-ventilated area away from windows and doors.

Do not use a generator indoors to prevent carbon monoxide poisoning.

MANAGING FOOD AND MEDICATION STORAGE

Refrigerated and Frozen Foods

Keep the refrigerator and freezer doors closed as much as possible to retain cool temperatures. Use perishable foods first and rely on non-perishable foods from your emergency supply kit.If the power outage is prolonged, consider alternative refrigeration options such as coolers with ice or seeking assistance from community resources.

Medications

Medication that needs refrigeration should be kept in a cooler with ice packs.

Contact your healthcare provider or pharmacist for guidance on medication storage during a power outage.

Temperature Monitoring

Use a food thermometer to ensure the refrigerator temperature remains below 40°F (4°C). Discard any perishable food that has been above this temperature for more than two hours.

Prescription Medication Refills

Refill prescriptions before a heat wave and power outage occurs to ensure an adequate supply. Consult with your healthcare provider or pharmacist regarding any concerns or alternative medication options during a power outage.

STAYING SAFE DURING POWER OUTAGES

Heat Wave Safety Zones

Identify local cooling centers, public buildings, or community facilities that offer air conditioning during a power outage. Seek shelter in these locations to escape extreme heat and reduce the risk of heat-related illnesses.

Heat-Relief Resources

Stay updated on local news, social media, or community websites for information on heat-relief resources, including emergency cooling shelters or community initiatives.

Stay Hydrated

Drink plenty of water and hydrating fluids to stay cool and maintain hydration levels.

Avoid consuming alcohol or caffeine-containing beverages in excess as they can cause dehydration.

Stay Cool Indoors

Create a cool room or designated cool area within your home using fans, open windows for ventilation, and utilizing shade. Wear lightweight, breathable clothing and use wet towels or cooling methods to lower body temperature.

SEEKING COMMUNITY SUPPORT AND RESOURCES

Community Assistance Programs

Stay informed about local community assistance programs that provide support during power outages and heat waves.

Contact local government agencies, non-profit organizations, or emergency management authorities for information on available resources.

Neighbor Check-Ins

Check on neighbors, particularly vulnerable individuals such as the elderly, those with medical conditions, or families with young children. Offer assistance and share information on available resources and support.

Community Support Networks

Connect with local community groups, neighborhood associations, or online forums to share information and resources during power outages and heat waves.

These networks can provide support, share updates, and offer assistance during challenging times.

By preparing in advance, managing food and medication storage, staying safe, and seeking community support, you can effectively cope with power outages during a heat wave. Remember to prioritize your safety, stay hydrated, and keep informed about available resources and assistance in your community.

8

EMERGENCY RESPONSE AND FIRST AID

During a heat wave, it is crucial to be prepared for potential emergencies and have basic first aid knowledge to respond effectively to heat-related illnesses and other health issues. Understanding emergency response procedures and knowing how to administer first aid can make a significant difference in ensuring the well-being and safety of yourself and others. Consider the following guidelines for emergency response and first aid during a heat wave:

RECOGNIZING HEAT-RELATED ILLNESSES

It is important to be able to recognize the signs and symptoms of heat-related illnesses. The following are common heat-related conditions and their associated symptoms:

Heat Cramps

Symptoms: Painful muscle cramps or spasms, typically in the legs or abdomen. Response: Move to a cool, shaded area, rest, and rehydrate with water or a sports drink containing electrolytes. Stretch or massage the impacted muscles gently.

Heat Exhaustion

Symptoms: Heavy sweating, weakness, fatigue, dizziness, headache, nausea, clammy skin, and rapid pulse.

Response: Move to a cool, shaded area. Loosen or remove tight clothing. Drink cool water or a sports drink. Apply cool, wet towels to the body. Seek medical attention if symptoms worsen or do not improve within 30 minutes.

Heat Stroke

Symptoms: High body temperature (above 104°F or 40°C), altered mental state, confusion, seizures, hot and dry skin, rapid breathing, and rapid pulse.

Response: Heat stroke is a medical emergency. Call emergency services immediately. Move the victim to a cool, shaded area while you wait for assistance. Remove excess clothing and use any available means to cool them down, such as applying cold water or ice packs to the neck, armpits, and groin.

EMERGENCY RESPONSE PROCEDURES

In the event of a heat-related emergency or any other medical emergency, follow these emergency response procedures:

Assess the Situation

Ensure your safety and that of others. Identify potential hazards or risks in the surrounding environment. Call emergency services or ask someone to make the call if professional help is required.

Provide Immediate Care

Move the person to a cool, shaded area away from direct sunlight and heat sources. Loosen or remove tight clothing to promote heat dissipation.

Offer water or a sports drink to rehydrate if the person is conscious and able to swallow.

Cool the Body

Use any available means to cool the person down, such as applying cool water or ice packs to the neck, armpits, and groin. Use fans or create airflow to aid in the cooling process.

Monitor Vital Signs

The person's breathing, pulse, and state of consciousness should all be monitored. Record and report any changes or deterioration in their condition to emergency services.

BASIC FIRST AID TECHNIQUES

In an emergency, having some basic first aid skills is vital. Here are some fundamental first aid techniques to be aware of:

Cardiopulmonary Resuscitation (CPR)

If a person is unconscious and not breathing or has no pulse, begin CPR immediately.Perform chest compressions at a rate of 100-120 compressions per minute and provide rescue breaths if trained to do so.

Recovery Position

Put a person in the recovery position to keep their airway open if they are breathing while unconscious and do not have any other serious injuries.

Wound Care

Cleanse minor cuts or abrasions with clean water and mild soap. Apply an antiseptic ointment and cover with a sterile dressing or bandage.

For severe bleeding, apply direct pressure to the wound and seek medical attention.

Heat-related Illness First Aid

Follow the appropriate first aid responses outlined in section 8.1 for heat cramps, heat exhaustion, and heat stroke.

FIRST AID TRAINING AND CERTIFICATION

To enhance your emergency response skills and knowledge of first aid techniques, consider enrolling in a certified first aid training course. These courses provide comprehensive instruction on basic life support, CPR, and other essential first aid skills. Having formal training ensures you can respond confidently and effectively in emergency situations.

Remember to always prioritize your safety and the safety of others when providing first aid. If in doubt or if the situation is life-threatening, call emergency services immediately.

By recognizing heat-related illnesses, following emergency response procedures, and having basic first aid knowledge, you can be better equipped to respond to emergencies and provide assistance during a heat wave or other critical situations.

9

LONG-TERM HEAT WAVE ADAPTATION

As heat waves become more frequent and intense due to climate change, it is essential to adapt and implement long-term strategies to mitigate the impacts and protect ourselves and our communities. Long-term heat wave adaptation involves implementing measures at both individual and community levels to reduce vulnerability and enhance resilience. Consider the following strategies for long-term heat wave adaptation:

INDIVIDUAL HEAT WAVE ADAPTATION

Home Insulation and Energy Efficiency

Improve home insulation to reduce heat gain and loss, making it easier to maintain a comfortable indoor temperature.

Install energy-efficient windows, seal gaps and cracks, and use reflective window films to block solar heat.

Landscaping and Shade

Plant trees and install shading devices such as awnings, pergolas, or shade sails to reduce heat around your home. Utilize natural shade from trees and use outdoor umbrellas or canopies to create cool outdoor spaces.

Cooling Strategies

Install or upgrade air conditioning systems to improve indoor comfort during heat waves. Use energy-efficient cooling methods like evaporative coolers, ceiling fans, or window fans to supplement air conditioning and reduce energy consumption.

Personal Protection

Wear lightweight, light-colored, and breathable clothing that provides sun protection during outdoor activities. Use sunscreen, wide-brimmed hats, and sunglasses to protect your skin and eyes from harmful UV radiation.

Hydration

Maintain good hydration by drinking water regularly, even when not feeling thirsty.

Carry a reusable water bottle to ensure access to clean water while reducing plastic waste.

COMMUNITY HEAT WAVE ADAPTATION

Urban Planning and Design

Incorporate green spaces, urban forests, and green infrastructure to mitigate heat island effects in cities. Plan for more parks, shade structures, and water features to provide cool and accessible public spaces during heat waves.

Emergency Response Planning

Develop comprehensive heat wave emergency response plans that include early warning systems, public health advisories, and cooling center strategies.

Enhance coordination among emergency management agencies, healthcare facilities, and community organizations to ensure efficient response during heat emergencies.

Community Engagement and Education

Conduct public awareness campaigns to educate individuals and communities about heat wave risks, prevention strategies, and adaptation measures. Promote community support networks, neighbor check-ins, and assistance programs for vulnerable populations during heat waves.

Green Infrastructure and Sustainable Practices

Encourage the use of green roofs, permeable surfaces, and rainwater harvesting systems to promote cooling and reduce stormwater runoff.

Implement sustainable practices such as renewable energy generation, energy-efficient buildings, and water conservation measures to reduce the environmental impacts of heat wave adaptation.

Heat-Resilient Infrastructure

Design and retrofit infrastructure to withstand extreme heat, including heat-resistant materials, heat-reflective surfaces, and improved ventilation systems. Incorporate climate change considerations in the design and planning of critical infrastructure such as hospitals, schools, and transportation systems.

COLLABORATION AND ADVOCACY

Collaboration

Foster collaboration among government agencies, community organizations, researchers, and stakeholders to develop and implement heat wave adaptation strategies.

Engage with local businesses, utilities, and industries to promote sustainable practices and reduce heat wave impacts.

Advocacy

Advocate for policies and regulations that prioritize heat wave adaptation and resilience in urban planning, building codes, and public health initiatives. Support initiatives to address climate change and reduce greenhouse gas emissions to mitigate the long-term impacts of heat waves.

By implementing long-term heat wave adaptation strategies at both individual and community levels, we can reduce vulnerability, enhance resilience, and minimize the health and societal impacts of heat waves. Adapting to a changing climate requires collective action, collaboration, and a commitment to sustainable practices for a more heat-resilient future.

10

DIY HEAT WAVE SURVIVAL MEASURES

When there is a heat wave, it can be difficult to deal with the extreme temperatures, especially if you don't have access to air conditioning or other cooling equipment. To withstand the heat wave and keep cool and comfortable, you can adopt a number of do-it-yourself precautions. You may significantly improve your ability to withstand the high heat by using these easy and affordable alternatives. Here are a few do-it-yourself tips for surviving a heat wave:

MAKE AN AIR COOLER AT HOME

Utilize a fan, an ice chest, or frozen water bottles to create a homemade air conditioner.

A cool wind can be produced by positioning the fan to blow air over the ice.

Make Your Own Air Cooler to Survive the Heat Wave

Keeping cool is crucial for your comfort and health during a heat wave. Although they are good at keeping indoor rooms cool, air conditioners can be expensive to run consistently. A DIY air cooler is a straightforward, economical solution that may offer cooling without consuming a lot of electricity. To survive the heat wave, try making your own DIY air cooler using the following steps:

Materials required:

- A big plastic bottle or cooler

- A cooler or container made of foam that is smaller than the large one

- Box fan, preferably, as an electric fan

- PVC pipe with a diameter of around 3 inches that is long enough to reach from the fan to the top of the big container

- Adapter for PVC pipes (to accommodate the fan)

- Either frozen water bottles or ice packs

- Cold water

Guide

- Prepare the big container:

- Take the large cooler or plastic container and take the lid off.

- Create a hole slightly smaller than the diameter of the PVC pipe on one side of the container. This will act as the fan's air inlet.

Prepare the small container:

Cut a hole slightly larger than the diameter of the PVC pipe in the lid of the smaller Styrofoam container. The outflow for the cooled air will be in this hole.

PVC pipe attachment:

Make sure the PVC pipe extends all the way to the bottom of the smaller container before inserting it through the lid's hole.

The PVC pipe adapter must be connected to the pipe's end and fastened securely.

Set up the fan:

Place the fan facing inwards next to the opening you made on the side of the big container.

Using the PVC pipe adapter, connect the PVC pipe's other end to the fan. Air will now be forced through the PVC pipe into the smaller container by the fan after being drawn from the large container's interior.

Add frozen water bottles or ice packs:

Leave space in the large container for the ice packs or frozen water bottles and fill it with cold water.

To further cool the water, submerge ice packs or frozen water bottles in it. The air that flows through the PVC pipe will become cooler thanks to the ice.

How to Assemble a DIY Air Cooler

Place the PVC pipe-attached lid over the smaller container.

Place the smaller container atop the larger one, making sure the PVC tubing is parallel to the air intake opening in the fan.

Activate the fan:

To start the airflow, turn on the fan. The fan will pull warm air from the vicinity, cool it with the chilled water and ice, and then release the cooled air through the PVC pipe into your living space.

Tips for Maintenance:

Regularly swap out the ice packs or frozen water bottles to keep the chilling effect going.

If necessary, add more cold water to the large container.

Maintaining a clean cooler will help to keep bacteria and mold from growing.

Remember that a homemade air cooler can offer great relief in small to medium-sized places even though it may not be as effective as a conventional air conditioner. It's an energy-efficient substitute that will allow you to weather the heat wave without breaking the bank. When utilizing the DIY air cooler, make sure your home has enough ventilation by cracking a few windows or doors to let hot air out and speed up the chilling process.

HOW TO MAKE YOUR OWN COOLING MISTERS

Use a spray bottle with water to make cooling misters. For a cooling effect, mix in a few drops of peppermint or eucalyptus essential oil. To instantly cool yourself, spray your body and face.

Making Your Own Homemade Cooling Misters

Using a cooling mister is an easy and efficient way to stay cool on a hot day. The water spray produces a fine mist that swiftly dissipates, instantly cooling your skin. Simple materials can be used to make your own cooling mister at home. Here is a step-by-step tutorial for creating your own cooling misters at home:

Materials required:

Obtain a spray bottle with a fine mist nozzle that is clean. Most drugstores carry empty spray bottles, and you may also buy them online.

Water: For your misting solution, use cool, clean water. You can use filtered water or water from the tap.

Optional: If you want a scented mist, think about using a few drops of essential oil.

Popular options for a reviving aroma include peppermint, eucalyptus, or lavender essential oils.

Guide

Clean a Spray Bottle:

To guarantee that the spray bottle is clean and residue-free, wash it first with soap and water.

Put Water in the Spray Bottle:

In order to prevent spillage, carefully pour cool water into the spray bottle after opening the nozzle.

Addition of Essential Oils:

Add a few drops of your preferred essential oil to the water in the spray bottle to create a fragrant mist.

Depending on your preferences and the scent's intensity, you may use more or less drops.

Snap the Spray Bottle Shut:

To avoid any leaks, firmly shut the spray bottle's nozzle.

Shake the Bottle:

Shake the bottle gently to combine the water and essential oil, if desired.

Examine the mist:

Keep your face and clothing away from the spray container as you hold it.

A thin mist will be released into the air by pumping the nozzle.

Verify that the mist is not excessively intense or weak. To achieve the required mist intensity, adjust the nozzle as necessary.

able to be used:

Use your homemade cooling mister at this point. On hot days, carry it with you at all times.

How to Use Your Cooling Misters

Maintain a distance of 6 to 12 inches between the spray container and your body and face.

To spray a fine mist onto your skin, pump the nozzle. Allow the mist to naturally evaporate so that it can chill your skin. When it's hot outside or during a heat wave, you can use the cooling mister whenever you feel warm or need to cool off.

Additional Advice

Before using, place your spray bottle in the refrigerator for a more energizing cooling mist. After each usage, drain any leftover water from the misting solution and replenish as necessary with fresh water.

It's inexpensive and environmentally friendly to make your own cooling misters at home to keep cool during the summer. In order to battle the heat and stay cool, a DIY cooling mister can be a useful companion whether you're lounging at home, working outside, or taking in a picnic.

USE BANDANAS OR WET TOWELS

Apply a towel or cloth that has been dampened with cool water to your wrists, neck, and forehead. This method can relieve discomfort and assist in regulating body temperature.

CREATE CROSS-VENTILATION:

To promote cross-ventilation in your home, open windows on opposing sides. To promote airflow, keep doors slightly ajar with door stoppers or wedges.

HOMEMADE SOLAR OVEN

Build a DIY solar oven to take advantage of the heat wave.

To cook food outside without using traditional heat sources, assemble a straightforward solar cooker from a box, aluminum foil, and a clear plastic top.

Making a DIY Solar Oven for Cooking During Heat Waves

The power of the sun may be used to prepare food without the use of traditional heat sources with the help of a DIY solar oven. Using a solar oven to prepare your meals

87

during a heat wave can help you avoid overheating your house with conventional cooking techniques. A fun and eco-friendly hobby that lets you cook great food while using less energy is building your own solar oven. Here's a how-to for building a basic DIY solar oven for cooking during a heat wave:

Materials required:

Box made of cardboard: Pick a box with a lid that is durable. Your solar oven's cooking capacity will depend on the size of the box.

A roll of heavy-duty aluminum foil can be used to line the interior of the box, reflecting light and preserving heat. Black construction paper: Cover the bottom of the box with this material. The dark surface will heat up as it absorbs sunlight.

A sheet of clear plastic or glass would work well to act as the oven's lid. While retaining heat within the oven, this material allows sunlight to travel through.

Gather insulating materials, such as newspaper, straw, or foam board, to line the box's sides to increase heat retention.

Cooking Rack: To set the food on inside the oven, you will need a cooking rack made of metal or wire.

The internal temperature of your solar oven can be checked with an oven thermometer.

Guide

Get the box ready:

Pick a box that will easily accommodate your cooking rack and food items.

Take off any tape or flaps from the box so that it is just an open container.

Utilize aluminum foil to line the box:

Aluminum foil should be used to line the interior of the box, shiny side facing inside. The foil will let light into the oven by reflecting it.

How to Make a Black Surface

Black construction paper should be used to cover the box's bottom. The sunlight will be absorbed by this dark surface and heated.

Add Insulation:

Insulate the edges of the box to hold heat in and increase the oven's effectiveness.

For an insulation layer, use straws, foam board, or crumpled newspaper.

Setting up the cooking rack:

Place the cooking rack within the container, leaving room for air to circulate around the food.

Shut the Lid:

The clear sheet of glass or plastic should be placed carefully on top of the box as the lid.

To close the lid and keep heat inside, use tape or glue to fasten it.

A solar oven test:

On a clear day, position your solar oven in the sunlight. For the most exposure, make sure the oven is tilted toward the sun.

To keep track of the oven's internal temperature, use an oven thermometer.

Begin cooking:

Place your meal on the cooking rack inside the oven once it has reached the proper temperature.

Put the lid on and let the sun's heat gradually cook your food.

Important Advice

Due to the possibility of a very hot interior, exercise caution when handling the solar oven and food.

To allow the most amount of sunshine to enter, make sure the transparent lid is spotless and free of any impediments.

To guarantee even and secure cooking, keep an eye on the oven's temperature frequently.

An eco-friendly and sustainable approach to prepare food during a heat wave is with your homemade solar oven. It's a great solution for low-heat cooking and making straightforward foods, even though it may not reach the same cooking temperatures as traditional ovens. Enjoy the process of using solar energy to prepare delectable meals while conserving energy in the summer.

HYDRATION STATIONS

Establish hydration stations with icy water, fruit-infused water, or drinks high in electrolytes. Keep reusable water pitchers or bottles full and accessible for each member of your family.

DIY SHADE STRUCTURES:

To offer shade in your outdoor settings, make your own shade structures out of bamboo, fabric, or other materials. Set aside cool locations so you may unwind and avoid the sun.

COOLING FOOT SOAK

Add a few drops of essential oils, such as peppermint or lavender, to a basin of cool water. To assist your body cool off, soak your feet in the cool water.

UTILIZE COOLING SLEEP AIDS

To create a cold sleeping space, briefly place your pillowcases and sheets in the freezer before bed. To improve your sleep during the heat wave, use cooling mattress toppers or gel pillows.

During a heat wave, keep in mind that your safety and wellbeing should come first. Seek emergency medical assistance if you or another person develops serious heat-related symptoms. While these do-it-yourself solutions can make you feel cooler and more at ease during a heat wave, it's crucial to know how to get to places that are cooler if the temperatures rise dangerously high. Take proactive measures to safeguard yourself and your loved ones during periods of excessive heat by being updated about heat advisories.

www.ingramcontent.com/pod-product-compliance
Lightning Source LLC
Chambersburg PA
CBHW062351290526
45794CB00005B/2172